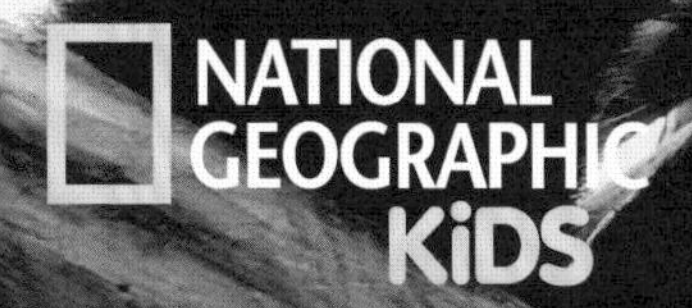

美国国家地理
双语阅读

Pandas

大熊猫

懿海文化 编著

马鸣 译

第三级

外语教学与研究出版社
FOREIGN LANGUAGE TEACHING AND RESEARCH PRESS
北京 BEIJING

京权图字：01-2021-5130

图书在版编目（CIP）数据

大熊猫 ：英文、汉文 / 懿海文化编著 ；马鸣译. -- 北京 ：外语教学与研究出版社，2021.11（2023.8 重印）
（美国国家地理双语阅读. 第三级）
书名原文：Pandas
ISBN 978-7-5213-3147-9

Ⅰ. ①大… Ⅱ. ①懿… ②马… Ⅲ. ①大熊猫－少儿读物－英、汉 Ⅳ. ①Q959.838-49

中国版本图书馆 CIP 数据核字 (2021) 第 226231 号

出 版 人　王　芳
策划编辑　许海峰　刘秀玲　姚　璐
责任编辑　姚　璐
责任校对　华　蕾
装帧设计　许　岚
出版发行　外语教学与研究出版社
社　　址　北京市西三环北路 19 号（100089）
网　　址　https://www.fltrp.com
印　　刷　天津海顺印业包装有限公司
开　　本　650×980　1/16
印　　张　37.5
版　　次　2022 年 3 月第 1 版　2023 年 8 月第 4 次印刷
书　　号　ISBN 978-7-5213-3147-9
定　　价　188.00 元（全 15 册）

如有图书采购需求，图书内容或印刷装订等问题，侵权、盗版书籍等线索，请拨打以下电话或关注官方服务号：
客服电话：400 898 7008
官方服务号：微信搜索并关注公众号“外研社官方服务号”
外研社购书网址：https://fltrp.tmall.com

物料号：331470001

Table of Contents

Look! Up in the tree!
Is it a cat? Is it a raccoon?
No! It's a **Giant Panda!**

Giant Pandas can climb to the tops of the tallest trees. They live in the highest mountains. They munch on bamboo for hours each day.

Bear Cat

Pandas are about the same size as their black bear cousins, but their heads are larger and rounder. Also, pandas cannot stand on their hind legs like other bears do.

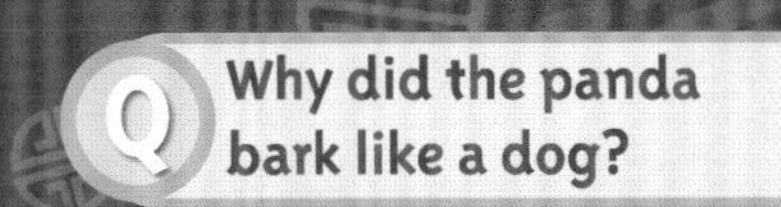

A
He couldn't BEAR to be
called a cat.

Pandas are a type of bear, but they seem more like raccoons or cats.

Like all bears, pandas are strong, intelligent animals with sharp teeth and a good sense of smell. Males weigh about 250 pounds and are about 4 to 6 feet long.

Black Bear

Pandas are great tree climbers. Sometimes they even sleep up in the treetops!

Bear Word

HABITAT: The place where a plant or animal naturally lives

Pandas have lived high in the mountains of China for millions of years. It is cold and rainy, but there are plenty of trees and a panda's favorite plant—bamboo.

Pandas used to live in more places, but today there is less open land with bamboo. Now pandas live in six forest habitats in China.

Panda Bodies

Pandas are black and white. This may help hide panda babies from predators, or enemies, in the snowy and rocky forests.

Their oily, woolly coat keeps them warm in the cold, wet forests where they live.

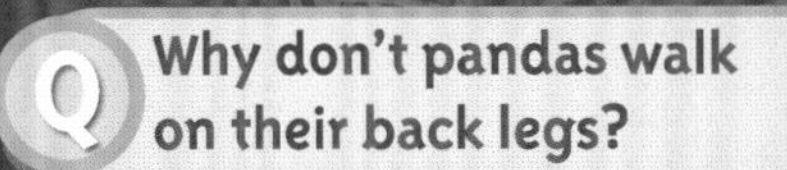

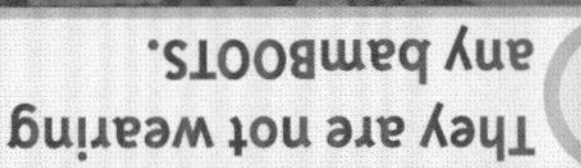

Their black eye spots may help them look fierce.

Just like cats, pandas can see very well at night, when they are most active.

Pandas have large teeth and strong jaw muscles that are perfect for crushing tough stalks of bamboo.

Bamboo Breakfast

Pandas spend their day sleeping a little and eating A LOT!

Bamboo for breakfast, bamboo for lunch, bamboo for dinner, and bamboo to munch. What do pandas eat? You guessed it—bamboo! It makes up almost all of a panda's diet.

Q Why do pandas like bamboo?

A It's panda-licious!

Pandas have to eat 20 to 40 pounds of bamboo each day to stay alive. It takes 10 to 16 hours a day to find and eat all that bamboo!

A Day in the Life

Q What do pandas eat on Halloween?

A Bamboo!

Pandas mostly live alone. But sometimes they hang out in small groups.

Pandas use 13 different calls to communicate with each other. They also leave their scent on rocks and trees for other pandas to find.

Baby Steps

Around August or September, a mother panda will find a den and give birth. Her newborn cub is about the same size and weight as an ice cream sandwich.

Panda cubs are pink, hairless, and blind at birth. They spend the day squeaking, crying, and drinking their mother's milk.

Soon, black fur will grow around the cub's eyes and on its ears and legs.

Cubs stay with their mothers until they are about two or three years old.

1.

In a few weeks, the mother can leave her cub to find bamboo. The baby cries less and is able to keep itself warm.

2.

When a cub is about eight weeks old, it will finally open its eyes. But the cub still cannot walk until it is about three months old.

What do you call a father bear?

A pandad!

3.

When the cub is about six months old, it can eat bamboo, climb trees, and walk around, just like its mother.

Red Panda

When people think of pandas, they are usually thinking of the Giant Panda. But did you know there is another kind?

Red Pandas also live in China as well as other parts of Asia. They eat bamboo just like black-and-white pandas, but they also love roots and acorns. Red Pandas only grow to be about the size of a cat.

Q What's black and white and red all over?

A A panda convention.

The Red Panda has red fur and looks more like a raccoon than a bear.

Protecting Pandas

Today there are only about 1,900 pandas left in the wild.

The Wolong (WOO-long) National Nature Reserve in China is just one way people are trying to help. The pandas that live there cannot be harmed.

Panda Baby Boom

Pandas are also protected in zoos. The first panda was brought to the United States from China in 1936. Today there are about 100 Red and Giant Pandas in zoos in the United States.

Q What do you get when you cross a know-it-all with a bear?

A A pan-duh!

In just one year, 16 cubs were born at the Wolong National Nature Reserve.

At first it was hard for Giant Panda moms to have cubs in zoos and on reserves. But in recent years, there has been a panda boom! Let's hear it for the cubs!

Earthquake!

In May 2008, a giant earthquake struck China. The center of the earthquake was right near the Wolong National Nature Reserve. Rocks the size of cars rained down from the steep mountains surrounding the pandas' home.

Panda-Mazing Facts!

Ancient Chinese rulers kept pandas as their **pets!**

Did you know?

Pandas will **roll** around and tumble to get somewhere faster.

Pandas are very **shy** and will stay away from places where people live.

Q: What do you get when you cross a playground with a bamboo forest?

A: Panda-monium!

Pandas are **pink** when they are born! The color comes from their mom's saliva when she licks them. (Saliva means spit!)

Pandas can't run very fast, but they are good **swimmers** and great **tree climbers.**

Pandas can eat more than **22,000 pounds** of bamboo each year!

It takes about **three years** to tell if a panda cub is a boy or girl.

Name That Bear

White Leopard

White Bear

Beast of Prey

Banded Bear

Pandas can be found in Chinese stories and poems 3,000 years old! Over time, they have been called many different things. Which name do you think fits them best?

Glossary

COMMUNICATE: To pass on information

CUB: A baby animal

EARTHQUAKE: When the Earth's crust moves, it causes the ground to shake.

HABITAT: The place where a plant or animal naturally lives

RESERVE: Protected land area

SCENT: A smell. Pandas use scent to communicate.

第 4—5 页

大熊猫！

看！在树上！它是猫吗？它是浣熊吗？不！它是大熊猫！

大熊猫能爬上参天大树的树顶。它们住在高山上。它们每天都要嚼好几个小时的竹子。

第 6—7 页

熊猫

大熊猫几乎和它们的近亲黑熊一样大，但是它们的脑袋更大、更圆。而且大熊猫也无法像别的熊那样用后腿站立。

黑熊

大熊猫是熊的一种，但是它们看起来更像浣熊或猫。

和所有的熊一样，大熊猫强壮，聪明，有锋利的牙齿和灵敏的嗅觉。雄性大熊猫大约 250 磅（约 113.4 千克）重，大约 4—6 英尺（约 1.22—1.83 米）长。

第 8—9 页

大熊猫在中国的高山上生活了数百万年。那里寒冷多雨，但是那里有很多树，还有大熊猫最爱的植物——竹子。

以前很多地方都生活着大熊猫，但如今有竹子的开阔地不多了。现在，大熊猫生活在中国的六个森林栖息地。

熊熊小词典

栖息地：植物或动物天然生长的地方

第 10—11 页

大熊猫的身体

第 12—13 页

竹子早餐

为了生存，大熊猫每天必须吃20—40磅（约9.07—18.14千克）竹子。找到并吃掉这些竹子每天要花10—16个小时！

大熊猫每天睡得很少，吃得很多！

早餐是竹子，午餐是竹子，晚餐还是竹子，不停地嚼竹子。大熊猫吃什么呢？你猜到了——竹子！它几乎构成了大熊猫的全部食物来源。

日常生活

大熊猫通常独自生活。但有时它们也会结成小群体出行。

大熊猫用 13 种不同的叫声来相互交流。它们也在石头上和树上留下自己的臭迹，方便别的大熊猫找到它们。

熊熊小词典

交流：传递信息

臭迹：一种气味。大熊猫用臭迹交流。

宝宝成长记

大约在八月份、九月份时，大熊猫妈妈会找一个兽穴生宝宝。刚出生的幼崽的大小和重量跟一块冰激凌三明治差不多。

大熊猫幼崽刚出生时是粉红色的，没有毛，也看不见东西。它们每天都在哭叫和喝妈妈的奶中度过。

不久之后，幼崽的眼睛周围、耳朵上和腿上就会长出黑色的软毛。

第 18—19 页

两三岁之前，幼崽都和妈妈待在一起。

1.

几周之后，妈妈就可以离开幼崽去找竹子了。宝宝哭的次数少了，也可以维持自身的体温了。

2.

大约八周大时，幼崽终于可以睁开眼睛了。但是幼崽还不会走，直到大约三个月大时才可以。

3.

大约六个月大时，幼崽可以像妈妈一样吃竹子、爬树、到处转悠。

第 20—21 页

小熊猫

人们一想到熊猫，总是会想到大熊猫。但是你知道还有另一种熊猫吗？

小熊猫也生活在中国和亚洲的其他一些地区。它们和黑白相间的大熊猫一样吃竹子，但是它们也喜欢吃根茎和橡子。小熊猫只能长到和猫差不多大。

小熊猫长着红色的毛皮，看起来更像浣熊，而不像熊。

第 22—23 页

保护大熊猫

如今，野外只剩下大约 1,900 只大熊猫。

中国的卧龙自然保护区只是人们尝试保护大熊猫的途径之一。生活在那里的大熊猫不会被伤害。

第 24—25 页

大熊猫宝宝潮

大熊猫在动物园里也可以受到保护。1936 年，第一只大熊猫从中国被带到美国。如今，美国的动物园里有大约 100 只小熊猫和大熊猫。

一开始，对大熊猫妈妈而言，在动物园和保护区生幼崽很艰难。但是，最近几年出现了大熊猫宝宝潮！让我们为幼崽鼓掌喝彩吧！

仅仅一年，卧龙自然保护区就出生了16只幼崽。

第 26—27 页

地震了！

2008 年 5 月，中国发生了一场大地震。震中正好在卧龙自然保护区附近。汽车大小的石头如大雨般从大熊猫之家周围陡峭的高山上落下来。

第 28—29 页

大熊猫趣事！

你知道吗?	中国古代的统治者把大熊猫当宠物养!	大熊猫刚出生时是粉红色的！这种颜色来自于妈妈舔它们时流出的唾液。（唾液就是口水！）	大熊猫跑不快，但它们是游泳好手和爬树高手。
为了快点儿到达某个地方，大熊猫会翻滚着前行。	大熊猫非常害羞，会远离人类居住的地方。	大熊猫每年能吃掉超过22,000磅（约9979.03千克）竹子!	大熊猫幼崽要在大约三岁以后才分得出是公的还是母的。

第 30—31 页

为这种熊起名

在中国 3,000 年前的传说和诗歌里可以发现大熊猫！长久以来，它们被取了很多不同的名字。你觉得哪个名字最适合它们呢?

猛氏兽

花熊

zzzzzzzzz
猫熊

熊猫

白狐

大熊猫

竹熊

第 32 页

词汇表

交流：传递信息

幼崽：动物宝宝

地震：地球板块移动使地面发生震动。

栖息地：植物或动物天然生长的地方

保护区：受到保护的区域

臭迹：一种气味。大熊猫用臭迹交流。